YOUR KNOWLEDGE HAS VALUE

AF154770

- We will publish your bachelor's and master's thesis, essays and papers

- Your own eBook and book - sold worldwide in all relevant shops

- Earn money with each sale

Upload your text at www.GRIN.com and publish for free

GRIN ☺

Purvesh Shah

Stereochemistry. An Introduction

GRIN Verlag

Bibliografische Information der Deutschen Nationalbibliothek:

Die Deutsche Bibliothek verzeichnet diese Publikation in der Deutschen National-
bibliografie; detaillierte bibliografische Daten sind im Internet über http://dnb.d-
nb.de/ abrufbar.

Imprint:

Copyright © 2012 GRIN Verlag GmbH
Druck und Bindung: Books on Demand GmbH, Norderstedt Germany
ISBN: 978-3-656-66079-8

This book at GRIN:

http://www.grin.com/en/e-book/272561/stereochemistry-an-introduction

GRIN - Your knowledge has value

Der GRIN Verlag publiziert seit 1998 wissenschaftliche Arbeiten von Studenten, Hochschullehrern und anderen Akademikern als eBook und gedrucktes Buch. Die Verlagswebsite www.grin.com ist die ideale Plattform zur Veröffentlichung von Hausarbeiten, Abschlussarbeiten, wissenschaftlichen Aufsätzen, Dissertationen und Fachbüchern.

Visit us on the internet:

http://www.grin.com/

http://www.facebook.com/grincom

http://www.twitter.com/grin_com

Stereochemistry- An Introduction

Dr. Purvesh J. Shah

Vanik Nivas, Ode-388210

Dist-ANAND,Gujarat(INDIA)

Stereochemistry-An Introduction

CONTENTS

Stereochemistry-An Introduction

AXIAL CHIRALITY

- The regular tetrahedron represent a three dimensional chiral centre.

- In which centre is occupied by tetra ordinate atom in which C is chiral centre.

- If this centre is replaced by linear grouping such as C-C or C=C=C, the tetrahedron becomes elongated along the axis as shown below.

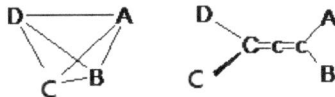

- Such an elongated tetrahedron has lesser symmetry than the regular tetrahedron and conditions for its dissymmertrisation is less.

- Here structure-II becomes three dimensionally chiral will give entionmer shown below.

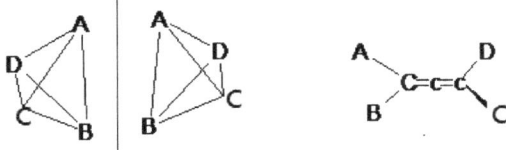

- The axis along which the tetrahedron is elongated shown by doted line is called chiral axis or stereoaxis.

- Some of the examples are:- (1) Allenes

(2) Biphenyls.

3

Stereochemistry-An Introduction

❖ **Allenes**:-

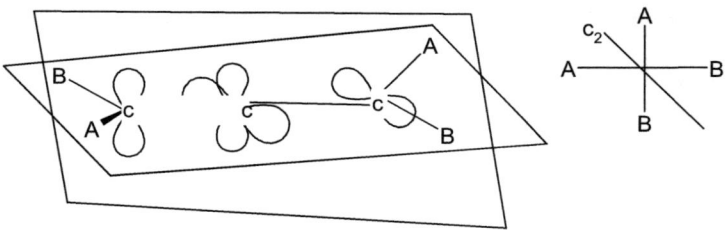

orbital picture of allene and its projected formula

- Here two end C1 and C3 in allenes are SP² hybridized and central carbon atom is SP- hybridized.

- In the orbital picture the shaded p-orbital and unshaded p-orbital separately overlape with each other forming π –Bonds making the two end groups non-planer.

- The structure can projected to Newman formula the allene posseses C2 axis, if 3 and 4 or substituents are different, the molecule is totally asymmetric.

- Here, first example 1,3-diphenyl 1,3-di-α-napthalenyl allene having R-configuration.

Stereochemistry-An Introduction

R

R

R

S

S

- •Spirens, alkylidenecycloalkanes and admentane appropriately substituted can be chiral in the same way as allenes.

- •Alkylidenecycloalkanes (hemispiranes) are compounds in which one of the double bond of allenes is replaced by ring shown in fig-1, and thus it will exhibit enantiomerism.

- ❖ **Optically active Spirans**:- The centre atom C-5 common to both the ring is chiral. Four group attached to C-5 of this compound is sequenced as C1>C6>C4>C9.

Stereochemistry-An Introduction

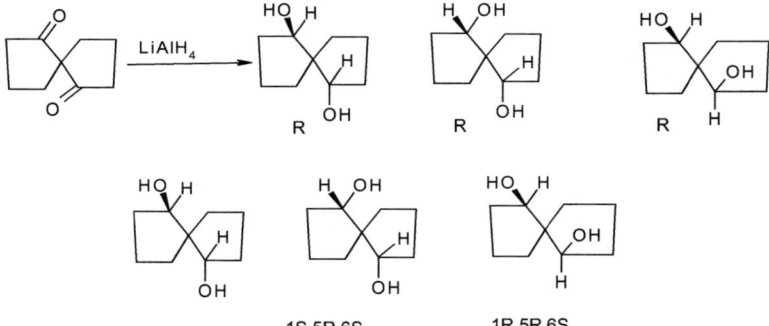

Spiro[4,4]-nonane-1,6-dione

- When molecule exhibit both central and axial chirality the formal has configurational nomenclature.
- When this diene is reduced to diol, three diastereomers diols are produced and each of these is resolvable.

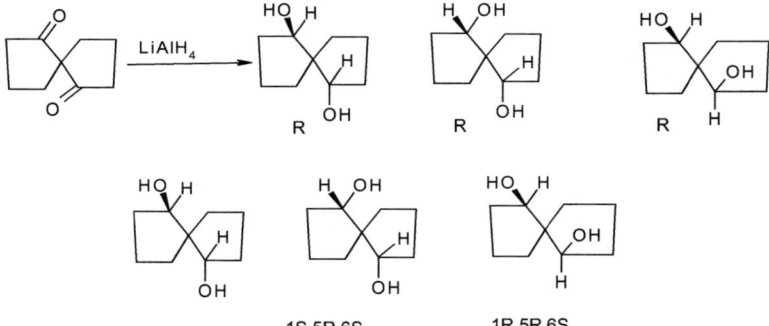

R   R   R

1S,5R,6S   1R,5R,6S

❖ Admantane:-

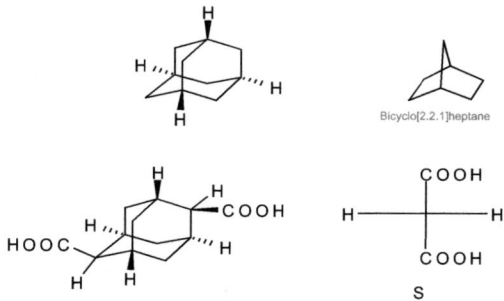

Bicyclo[2.2.1]heptane

admantane-2,6-dicarboxylic acid

S

6

Stereochemistry-An Introduction

- It satistify the condition of axial chirality in which C2 and C6 methylene are non-planer and in two enantiomeric form.

- Chiral axis passes through C1 and C6.

❖ Catenanes (optically active):-

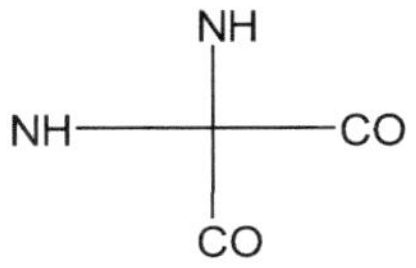

- A catenan with two dissimilar rings interlink with each other may give rise to chirality due to secondary structure.

- If the two rings are held with their planes perpendicular to each other in a following structure with 4 different groups in the chain.

- Like other axially chiral molecule configurational nomenclature will be given from the projection formula.

- Thus particular enantiomer has "R" configuration.

❖ Biphenyl:-

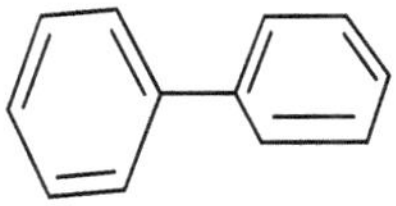

- Biphenyl is represented by the structure in which two phenyl groups are joined by single bond is called pivotal bond. Configuration of both carbons is SP^2.

- The distance between ortho hydrogen in planer conformation is 0.29nm which is greater than Vander walls radius of Hydrogen. i.e.

Stereochemistry-An Introduction

2*0.12=0.24nm. So rotation about the pivotal bond is not impeded by stearic factor.

- The phenyl group which is dissymmetrically substituted is two dimensionally chiral and affords cis and trans diastereomer. So, this is planar combination. If a non-planer combination would give enantiomer and such molecules are axially chiral. Thus, biphenyl has resolved properties.

- This can be done by introducing bulky groups in the ortho positions, so the planar conformations are dis-stabilised by stearic repulsion.

- The energy diagram is shown in fig-a.

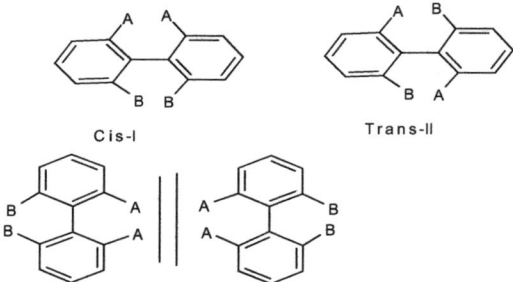

❖ Essential conditions of chirality of biphenyls:-

- Biphenyls with 4 bulky groups in ortho position are not free to rotate about the central bond. So, two rings are must be in perpendicular plane with each other.

- When either or both the rings are symmetrical molecule will have plane of symmetry and will be Achiral.

Stereochemistry-An Introduction

- When neither ring is symmetric compound is chiral as shown in structure.

R

- It is not always necessary that 4 large ortho group should be present for rotation to be prevented. The presence of lesser no. of group, if large enough can prevent rotation which exist enantiomer in the following case.

S

- The hydrogen-atom is quite small optically active compound exist with 2 or 3 ortho positions of the biphenyls occupied by hydrogens. In this case benzene ring is unsubstitued at ortho position but substituted at meta position, m-substitution has no influence on rotation. But it creates necessary chirality in the molecule as in "3-bromo-biphenyl-2-trimethyl Aarsonium iodide."

Stereochemistry-An Introduction

- In this case single bulky group prevent rotation to make the compound chiral.

- In the several cases the groups may be large enough that extent this slow to rotation but completely preventing. Such chiral compounds slowly recemizes on standing shown below.

- The enantiomers are inte4rconverted involving squizing of the smaller 'F' pass the adjacent carboxylic group via plannar conformation.

- Once a planar conformation is reached, chirality is lost to result in recemization.

Stereochemistry-An Introduction

ATROPISOMERS

- Two conformation of molecule whose interconversion is sufficiently slow under given set of condition to all their separation, these are known as <u>Atropisomers</u>.

- They are also known as "tortional isomers" about single bond.

- They can't be represented by any kind of formula as in the case of allenes, pyranes andother compound.

- Physical study, X-ray diffraction, dipole moment measurement, electron diffraction study shows the non-planar conformation.

1

R

2

S

3

Stereochemistry-An Introduction

- First two are resolved but (3) in which positions are joined by planar ring is non-resolved because non-bonded interaction has been converted in to bonded one.

- Dinitro-diphenic acid was first optically active biphenyl which was resolved.

- The 'R' enantiomer is show above.

1   R

O_2N—⎸—$COOH$
$COOH$ / NO_2

2

3

4

5

Stereochemistry-An Introduction

- Dinitro-diphenic acid compound-1 is resolved this basically depends on the sum of van der walls group radii of ortho substituents.

- This is illustrated in compound-2 in this case carboxylic group serve two purposes. They make the phenyl ring dissymmetric and two dimensionally chiral centres provide a handle for resolution.

- If R is fluorine, the radius is 0.278nm (0.139+0.139=0.278nm) is smaller than the require overlapping. Thus the compound-2, if R is fluorine is not resolvable.

- If R is –OMe group **overlapping** occurs and compound recemizes. Radius is 0.149 +0.149=0.29nm.

- On other hand R=Cl than 0.169+0.169=0.34nm. So, it is considerable overlapping takes place. So, two enantiomers can be separated.

- Thus depending on ortho substituent biphenyl can exist in stereoisomers which range from flitting to stable configurational isomer.

- The order of stearic hindrance produce from various group appear to be

Br>Me>Cl>NO2>COOH>Ome>F.

- Compound-4is duetrated and is 1.13times faster than unduetrated comp-3.

- This is known as secondary isotope effect, because duetrarium has smaller Van der walls radius than hydrogen as its lower 0.5 vibrational frequency.

Stereochemistry-An Introduction

- Compound-5 4, 4',6, 6'-hexahydroxy diphenic acid. It is obtained naturally and it is optically active.

- Biphenyl atropisomers with only two bulky ortho substituent are also known.

2,2'-DITERTIRARY BIPHENY

- It can also give atropisomers, although it contains three Hydrogen substituent at ortho position.

- The nature and position of other substituent in ring play role to determine configurational stability to atropisomers.

- It is due to three factor:-

    A bulky group adjacent to a ortho substituent affect buttressing effect. Thus the rate of recemization 3'-Nitro derivative of following compound is much lower than 5'-nitro derivative compound.

Stereochemistry-An Introduction

- This effect of some of the groups are in order –NO2>-Br>-Cl>-Me

- Which does not exactly corresponds to stearic hinderence.

- A group at 4 and 4' position retard the recemization that is due to change in entropy of activation.

- The presence of electron attracting and electron-donate at 4 and 4' positions help interannular resonance to the pivotal bond. Thus transition energy is lower but effect is found to be much smaller.

- Atropisomerism is not limited to biphenyl derivative; it is also encounter to Binapthyl, Bipyridyl and Bipyrrole.

- It may also form by the compound in which benzene ring is replaced by ethylene group.

Stereochemistry-An Introduction

[Chemical structures shown: various naphthalene and pyridine biphenyl-type compounds with COOH groups, including a binaphthyl dicarboxylic acid, a bipyridyl dicarboxylic acid, an N-aryl pyrrole dicarboxylic acid, an N-N linked pyrrole carboxylic acid, and an O₂N/Ph-SO₂ substituted naphthalene with CH₂COOH.]

❖ The Atropisomerism of 2,2'-bridge biphenyls:-

There are three types of bridge biphenyls are as follows.

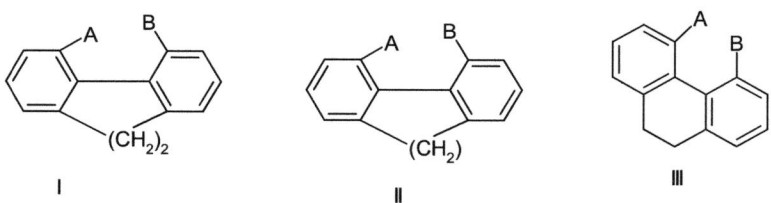

- When a biphenyl 2,2'-position are bridge as shown in one(I)with the rings of different sizes. It can consider three situations.

- When n=1 the compound is plannar disubstituted (II).

  n=2 compound is dihydrophenanthrene (III).

Stereochemistry-An Introduction

- Thus non-plannar six membered rings can lead to atropisomers provided the other two ortho positions are substituted with bulky groups.

- No. if n>2 will the bridge display Atropisomerism irrespective of the size. So, in this situation non-planarity is forced by puckering of the ring due to non-bonded interaction and angle strain. In reference to this examples are given as below,

- This bridge biphenyls recemizes with the relative because angle strain and stearic interaction in the medium rings are not very large.

- They can be further stabilized by substituting the remaining ortho position with bulky group as in keto (A).

- Ex. (B)-thio ether –can be resolved because, both the ring are purked.

❖Configurationally nomclature of biphenyls:-

Stereochemistry-An Introduction

- The molecule is viewed from either end of the chiral axis and projection formula is drawn. Only C2 and C6, C2' and C6' which corresponds to four vertical of the elongated tetrahedron are considered.

- If the viewed from the left side 1-1'bond will give the projection-A and when viewed from the right hand side 1'-1 bond give the projection-B. Looking to the figure, both the projection confirmed as the configuration.

- 2,2'-bridge biphenyls can be treated similarly.

❖ Atropisomers other than biphenyls:-

- One of both of the phenyl ring are replaced by other aromatic or heterocycles. Thus substituted N-phenyl pyrol, N,N'-biphenyl pyral, 1,1'-binaphthyl-3.3'-bipyridyl are resolvable.

- Two phenyl rings in biphenyl are interposed by a phenylene ring forming p-terphenyl derivatives and restricted rotation may arise around two pivotal bonds.

- So, the two phenyl rings are co-axial or coplanar. This corresponds to a planar combination of two-two dimensional chiral units and both diastereomerism and enantiomerism may results.

- The, terphenyl derivatives shown in figure.

18

Stereochemistry-An Introduction

C i s

T r a n s

• Here, all eight ortho positions are substituted exist in two achiral diastereomers. This can be described as cis and Trans (with reference to Br atom).

• In cis-from:-central ring is dissymmetric. Thus it forms enantiomerism.

• In trans-form:- there is a centre of symmetry.

• One of the planar rings is replaced by acyclic group which is two dimensionally chiral usually substituted trigonal atoms. It gives the Atropisomerism if sufficient stearic hindrance created around the pivotal-bond.

• So, in first compound substitution of stelebene is capable of resolution.

• In second (2) the derivative is naphthylamine in which peri-nitro group prevent the substitution at nitrogen to cross the plane of naphthalene ring gives the "s" conformer.

Stereochemistry-An Introduction

• 1-acetyl-2-hydroxy napthlene-3-carboxylic acid forms two oxime, one is cis and other is trans.

• They are two diastereomers; in first rotation around aryl carbon is not sufficiently restricted. So, the compound is non-resolvable.

• In II hydroxyl group of oxime in the planar conformation interfere with the substituted which permit the resolution.

• Upon treating both with hydroxylamine hydrochloride (Beckmann rearrangement) gives different products.

❖ Atropisomerism around SP³-SP³ bond:-

• Atropisomerism is also due to restriction rotation around a sp²-sp² single bond.

• It is also known the rotation about sp3-sp3 single bond, is restricted to some extent because energy barrier is very low.

Stereochemistry-An Introduction

- In the molecule triptycene the barrier to rotation around 9-substituted bond is quit high, thus conformers given below.

- The first give the meso isomer, while str.(II)and (III) are two enantiomer. They are stable at room temp.

I

II

Mirror

III

Stereochemistry-An Introduction

❖PLANAR CHIRALITY:-

- The geometrical requirement for the planar chirality is exchange the ligand across the chiral plane which gives the enantiomers.

- Configurational stability arise due to restriction impose by stearic factor on the out of plane movement of a structural unit. The energy factor is also considered. This type of the molecules is not represented by any formula because stereoisomer isolated base on bulk of groups and ring.

- This is also type of Atropisomerism and it is exhibited in following compounds,

  (1) Ansa compound

  (2) Cyclophanes

  (3) Trans-cycloalkenes.

(1) Ansa compound:- if two p-position of aromatic ring are attached to heteroatom and they are in turn connected through polymethylene chain, the compound are called Ansa compound.

Stereochemistry-An Introduction

- Here all of these the aromatic ring is dis-symmertrically substituted and the polymethylene chain can be either above or below, the plane of aromatic ring giving two enantiomeric structures.

- Now, if the polymethylene chain is small enough it can not be swing around the plane and two enantiomer will be configurationally stable.

- Thus, compound-1 with monosubstituted phenyl ring is resolved, when n=8. Recemizes easily when n=9.but is non-resolvable when=10.

- Disubstitued compound-2 gives stable enantiomer, even when n=10.

- Compound-3 is extremely stable.

- For the assignment of chirality in these compounds-a pilot atom is to be selected first which is directly bonded to atom in chiral plane but itself is not in the plane. So, the pilot atom should be chosen side of the plane.

- Thus in the case of first the preferred side of the ring (1) with ortho bromine and left hand methylene carbon is pilot atom.

- In case both side equivalent the pilot atom may be chosen from any side compound (2).

- The sequence stars from first in-plane atom (hetero-atom) and continue through atom in the plane.

- Thus, if order of rotation will be clockwise configuration will be "R" and if it is anticlockwise than it will be "S".

(ii) Cyclophanes:- Cyclophanes are similar to Ansa compounds. Here, usually two aromatic rings are joined together by bridging p-position to give p-Cyclophanes or meta position to give m-Cyclophanes.

Stereochemistry-An Introduction

- Compound-1 is simple Cyclophanes with one aromatic ring, only one which have been resolved, it resembles to Ansa-compound.

- Here, pilot atom is methylene carbon on the side of carboxylic group and stru-1 has "R" configuration.

- Compound-2 , here two rings are arranged one above and other parallel and one ring with substituted carboxylic group can't make full turn if chains are small.

- Compound-3, one ring is dissymmetrical and will exist enantiomer.

- Paracyclophanes, two parallel aromatic π-electron systems interact with each other, called **π-π -Trans annular effect.**

(iii) Trans-cycloalkenes:- Trans-cycloalkenes will provide another type of molecule with planar chirality. Two trigonal carbons and atoms directly attached to them are in the plane and polymethylene bridge squewed in third dimension.

Stereochemistry-An Introduction

e.g. cyclooctane-it is smallest ring which can accommodate a trans-double bond and two conformations are possible, which are mirror image of each other.

Here A and B are mirror images of each other.

- The interconversion of two enantiomer which require the swinging of tetramethylene chain over and below the plane of trigonal atom is opposed by ring-strain and two enantiomers are to be separated.

- Trans-cyclononene exists in optically active form at -80°C and Trans-cyclodecene is extremely mobile phase.

- In ferrocine- one of the five membered ring dis-symmertrically substituted. If it is a typical example of chiral metallocene, which is considered having planar chirality.

- In molecule each of the 5-carbon atoms of substituted cyclopentadienenyl ring is an asymmetric atom and bonded to the metal.

- The chirality of centers, C_1-Fe (a) C_2-(b)

                                              ▪C_3-(c)   COOH-(d)

- It has "R" configuration.

Stereochemistry-An Introduction

HELICAL CHIRALITY

- Helix is chiral objects and thus a helix is non-superimposable on its mirror image.

- The assignment of chiral designation to a helix is straight forward a right handed helix (clockwise) is designated as 'P' or '(+)' and left handed helix (anticlockwise) is designated as 'M' or '(-)'.

- At the molecule level helicity arise due to 2°-structure and is conformational in origin.

- Helical structure has been extensively studied in protein molecule.

- Intramolecular over crowding may leads to helicity in molecules and hexahelicine is one of the good examples.

hexahelical

P

M

- Through the molecule is normally expected to be the planar terminal benzene ring in hexa-helicene can't occupy the same plane without coming in contact each other.

26

Stereochemistry-An Introduction

- Thus, molecule is force to adopt a non-planar shape in which one side of molecule must lie above the other because of crowding. Molecular-overcrowding can also be introduced by suitable substitution in phenanthrene and benzphenanthrene.

- Phenanthrene is a planar molecule but when substituted at position (4) and (5). e.g. –Me group it becomes skewed and exists in two enantiomers. So, it is easy to isolated two enantiomers shown in fig.

- Conformational helicity is also encountered in benzophenanthrene in which again the terminal ring and substituents are in different planes and it exists as two helical enantiomers.

❖ Criteria for chirality:-

- Each plane in this compound is one naphthalene unit the angle between two planes containing terminal benzene ring is around 58.5°. Thus in hexahelicene, the middle rings are in a plane and terminal rings are one fall above and other below the plane respectively.

- Hexa-helicene is chiral by virtue of its helical shape, which could be either left or right in orientation.

- The entire molecule is less than one full turn but enough to generate chirality. Both, the enantiomers are stable and separated or resolvable.

Stereochemistry-An Introduction

<u>CYCLOSTEROISOMERISM</u>

- New type of the stereoisomerism is based on cyclic directions compounds exhibiting such phenomenon. These have cyclic structure and have more than one chiral centre either as a part of ring or a side chain. One of the simple example is 2,5-diketopiperazine and it is undirected, if they are equivalent as benzene and cyclohexane.

- The necessary and sufficient condition of cyclic directionally compounds is the absence of C2n-axis and 6-plane bisecting the molecule.

- If two such molecules have identical frame work but differ in ring directionality they are further divided into cycloenantionmers and cyclodiastereomers depending on whether they are mirror images of each other or not.

e.g.

Ala-Ala

- Cyclolligo peptide provides example of ring compounds with cyclic direction. If such cyclo olligo peptides is built up of equal no. of R and S amino acid. It may show cyclosteroisomerism.

Stereochemistry-An Introduction

•When total no. of Alanine residue is two then only one arrangement is possible.

•A short hand notation for a two unit chain containing one R and one S alanine moiety shown in figure to be used in next discussion. White circle stands for 'R' and cross circle stands for 'S' chiral centre. The arrow (⟶ pointing to right and arrow (⟵ pointing to left stands for NHCO and CONH respectively.

•When residue are- 4(four), two different arrangements of chiral centre.

   (1)RRSS   (2) RSRS are possible.

•Both of them are achiral (meso compound) as both are non-resolved due to meso form. Both are shown structurally as follows:-

•When n=6, means six amino acids are linked together three different arrangement of chiral centre position are possible:-RRRSSS, RSRSRS.

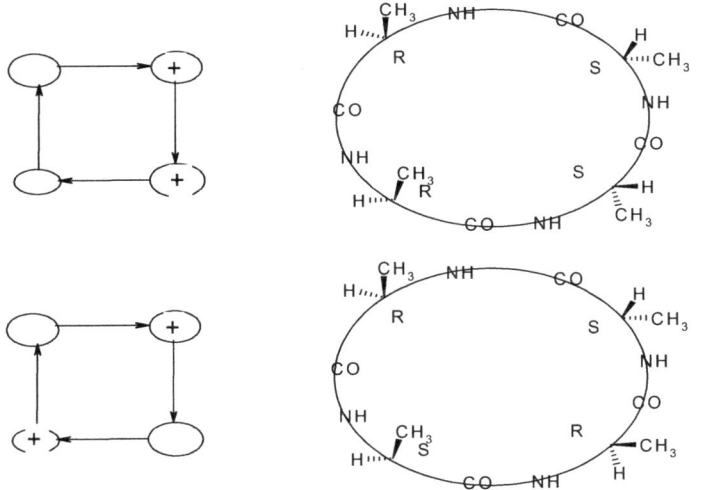

(1) The one in which 3-R and 3-S centers are placed consecutively and form meso-isomers.

Stereochemistry-An Introduction

(2) In second –R and S placed alternatively it is also meso compound and Its mirror image is superimposable.

(3) In third- it does not have any S_n-axis and is chiral and these are called cycloenantionmer.

- All the R and S units are arranged exactly as in the original, only the ring directionally different which prevent two forms being superimposable; these structures are called "**cycloenantiomers**".

- When Amino acids are 10 (n=5).Out of 10,3 occurs meso meso forms, six pairs of cycloenantiomers and five pairs of enantiomers.

- Following structures (i) and (ii) represent a normal enantiomeric pair, while the structure (iii) ,a member of another enantiomeric pair.

- Examination of 1st and 3rd reveals that their chiral frame works are identical, only ring directionality is different. Thus they cannot be in cycloenantiomer due to not mirror images but they are called cyclodiastereomers.

  1. The enantiomer of 3rd is in turn of cyclodiastereomeric.

  2. The characterization of pair cyclostereoisomers has three basic criteria.

a. Identity or non-identity of chiral frame work.

b. ring directionality

c. Mirror image relation

Stereochemistry-An Introduction

RETEROENANTIOISOMERISM

- Cyclostereoisomerism having two or more chiral centre gives this type of isomerism. If the configuration of each chiral unit and ring direction are both reverse for a structure, a new structure results which is an isomer of original in which the sequence C_1-NHCO-C_2 is replaced by C_1-CONH-C_2.

- C_1 and C_2 represent two different chiral centers. Two such dipeptides are shown in figure. The circles of different radii referred to different amino acid residues. Such pair of compounds are called Reteroenantioisomers.

- They are of biological interest because one can effectively replaced the other in biological reactions.

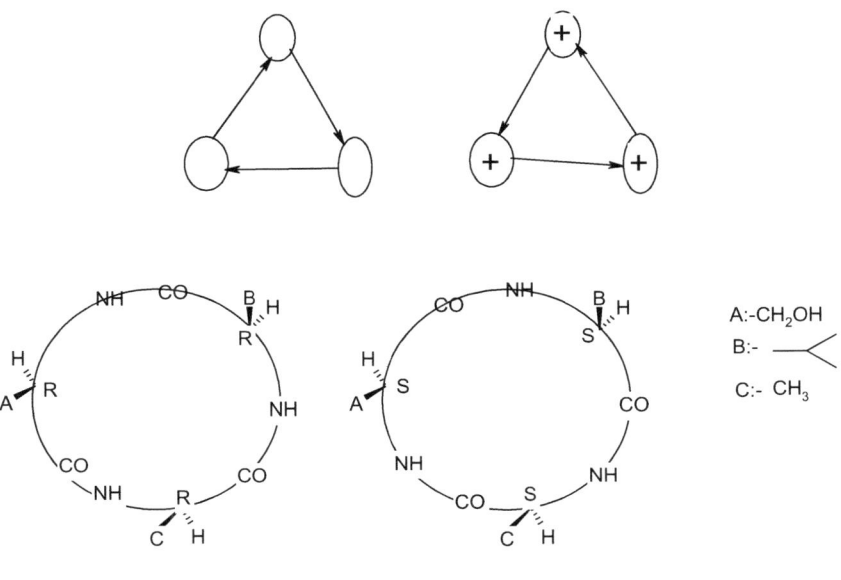

++